Title

Code Brown is the term used by the medical profession to alert nurses that someone has soiled themselves and needs to be cleaned up.

Dedication

To My Parents:

Lorraine A. and Oliver J. Domingue, Jr.

The bravest pair I've ever known.

Copyright 2012: Terry Derwitsch

#1-784908888

Publisher: Lulu.com

Introduction

Americans in 2012 have found themselves in the worst medical climate possibly in the history of man. We all saw the problem coming, too. It was many years ago, that presidential administrations first started promising to fix it.

But something happens when they come into office. They learn how powerful the medical and insurance industries are. They are so powerful that currently, they are almost impervious to any outside influence. These two industries have so much money and control; they could advocate the removal of tongues in adolescence for talking on the phone too much and get away with it.

These groups control health care in our country. They and only they decide what the proper treatment for a patient is, what will be covered by insurance and not covered by insurance, how everything will be complicated beyond belief and how much it will all cost.

How much does it cost? Way too much. You've heard the stories. Seniors choosing between food and medicine, disadvantaged kids going without proper medical or dental care, folks who can't afford the treatment they need to stay alive. Conversely, people of means or patients with good healthcare are getting way more treatment than they need or want.

The testing practices and drugs that are peddled are scary and dangerous and often unnecessary. These various pre-disease treatments are designed to scare the living daylights out of people and force them to choose a course of action that will drain their bank account, put their health in danger, (generate medical records making them ineligible for insurance), and most likely won't improve their imagined future health problems.

The American public has to shoulder some of the responsibility for how this happened. In my research for this book, I found that an alarming number of people have no idea how their bodies really work. Also, they couldn't tell me exactly about the treatment they were receiving or even the names of their prescriptions and/or diseases! It's not surprising that medicine is taking advantage of people when they are not the least bit interested in taking control of their health treatments or asking questions or making their own decisions.

Can it be fixed? I'm certain that it can. I'm certain because everything can be fixed or at least, greatly improved. But it's going to take an enormous shift in thinking and involvement from American citizens to make it happen. Ready?

Essays

One: The Grim Reaper
Two: False Hope is Not Hope
Three: The Blame Game
Four: Testing… 1, 2, 3, Infinity
Five: The Biggest Scam: Insurance
Six: Sueage
Seven: Records Breaking
Eight: Hereditary Risks
Nine: Newfangled Diagnoses
Ten: The Cancer Circus
Eleven: Fake Pretty
Twelve: The Real Drug Problem
Afterward

The Grim Reaper

About 15 years ago, I was talking to a friend who was 85 years old. Her husband was ill and she said to me, "I hope nothing happens to him." My heart broke for her; they were married for more than 60 years. But my mind thought her line of thinking was way off base. Of course something is going to happen. He's 87!

I'm sorry to be so crude but, if you want to handle the death of your loved ones and your own death in the best way possible, it's essential that you accept death first. How do I know this is true?

When I was 11, I was told that my 8 year old brother and my newborn baby brother had muscular dystrophy. The baby had a myriad of other problems, too. Being the wide-eyed learner I've always been, I sought all the information I could about this awful disease. So, I knew when I was 11 years old that both of my brothers were going to likely die during their childhoods.

And I met literally thousands of other kids with lots of other untreatable diseases who were living perfectly happy and adjusted lives despite the fact that early death was coming; and there was no way to get out of it. No medicine, no treatments, we all just had to accept it and live with it.

If I ever forgot about it for a minute, the public reminded me. This was pre-pre-Americans with Disabilities Act—it was still okay to stare at and shun crippled people and their families. Don't go crazy on me now because I said crippled. I was told once by a 16 year old friend with cerebral palsy, "So many friends have inadvertently peed on you; you have the right to say anything!"

And Jerry Lewis reminded me every fall. (Thank you, Jerry. I love you. You could never imagine how much you meant to us!)

I didn't know it at the time, but I started preparing for these death events, the minute I learned of them, in my mind and heart, and through my experiences.

Fortunately, we had parents that were highly enterprising, crazy strong and fantastically supportive. They made as happy and normal a childhood for us as they could. They were people of modest means, but they still belonged to all the associations helping other families. In spite of muscular dystrophy, I had so much fun and was so loved that I didn't even know I was poor until I grew up! I had bigger problems to deal with than money.

When I was in high school, a well-loved boy in my class was killed in a mini-bike accident in our town. I was able to attend his funeral with composure and help his younger sister face her fears of praying beside his casket, when I found her in the sitting room crying. How did I do this as a teenager? I was already getting used to death for a few years by then.

When my brothers did die, (ages 21 and 16 within a month of each other) I was 24 and I was ready for it. Was I angry? Yes. Was I devastated? I most certainly was. But nowhere near as much as I would have been had I not been preparing myself for it to happen.

This was a great gift to me that has helped me many times throughout my life and especially at that time. There was no one left but me to support and help my distraught parents get through our family's tragedy. They had no idea how to live without those boys. They had to learn how to do it.

Of course we had tons of friends and family to surround and love us, but most of them were just as distraught as my poor worn-out parents. I was able to comfort them all. When they asked me how I could be so strong, I said, "We don't have any choice." It's absolutely true. We don't. We are all going, folks. For real. My brothers are long gone and my Daddy's gone now, too.

I do not embrace death. And I'm not saying you have to be a fan of death. I'm saying you have to know death is coming and let go of your fear of it. If you don't, you will be unable to make smart decisions about your health care or anything in your life for that matter. This is the single most important thing YOU can do without cost to fix healthcare in America.

Doctors and pharmaceutical companies can terrorize you into paying them all the money you have and spending all your remaining time on earth in complete distress--all in vain.

In lots of cases, there is no medical treatment that can stop death. I know firsthand that acceptance of death makes a big and positive difference in peoples' health care choices and their lives. There are tons of places you can get more information about these subjects to help you accomplish this goal. I would suggest that you start with Dr. Elisabeth Kubler-Ross. Her books were of enormous help to me.

Of course your own personal experiences will help you the most. For this reason, take every opportunity to pay extra attention and time to those you know who are struggling through these events of death. You will know you've done something important for someone in real need. You will feel it. Then you will benefit in personal growth toward the goal of facing your own death, unafraid, with a clear and accurate mind if at all possible.

False Hope is Not Hope

I'm proud to say I'm the enemy of false hope. Sometimes I'm misunderstood as cynical because of this. I'm not advocating that people give up hope. I would never do that. In fact, I'd say I am as hopeful as they come.

I'm advocating that people give up false hope. It is amazing how many patients I have met in my research that never learn anything about their conditions and blindly follow any doctor's advice.

I'm not a doctor! I'm a patient. Let me be clear: the advice I am giving you is not medical advice. It is patient advice.

This means I want you to be sure that you know every detail of your anticipated treatments and prescriptions. I want you to know how your bodies work. You should know what your alternatives are. You should also know where the treatments came from and why, what they involve and what the risks are. Look for the numbers on positive results and negative results of the treatments. Use math to figure out what it all means. Be honest with yourselves about your circumstances. If you've already decided that you don't want the treatment; just say no.

Know that many states, including New York, Florida, and New Jersey have provisions in their Constitutions to prevent anyone from stopping you or mistreating you if you want to refuse their suggested treatments. There are several groups working to assure that the US Constitution also has an amendment for this purpose. That is how very important your choices are!

Use your new way of thinking about death without fear along with all of the factors mentioned above combined. Then, at a time when your emotions are in check, you make a decision that feels right for you. Your body belongs to YOU. It is your personal temple— you decide what the course of action should be. It's going to be different for everyone because we are all so different.

One more thing you can do to help us all: If your doctor drops you as a patient for refusing to go along with a suggested treatment, report him to the state authorities where you live. This will help to curb the bullying of patients to adhere to lots of expensive treatments and dangerous drugs that are most likely accomplishing little more than making the drug and medical industries richer.

You must have learned this is true in your research if you did refuse some sort of treatment. Stop these physicians from taking advantage of patients who are less informed or less brave than you are—report them to the proper authorities if they will no longer care for you simply because you disagreed with them. This is unethical; it's against the constitution of many states, and it has to be stopped in order to change healthcare.

The Blame Game

Somewhere in the middle of my life, it became common practice to blame people for their own diseases. I think it was probably part of the browbeating tactics adopted when we began trying to pressure everyone into quitting smoking, once we had everyone hooked on nicotine.

This hideous practice must be discontinued. Not only is it hurtful and unnecessary to blame people for their illnesses, in the majority of cases, it is pure bunk.

Here's how we know: Jim Fixx.

Jim Fixx is the man who is credited with beginning the fitness craze in the 1970s that is still raging today. He wrote the bestseller, *The Complete Book of Running* which popularized running for fitness and touted the benefits of regular jogging. It sold over a million copies. In the seventies, that was a lot of copies. He also wrote many other successful fitness books.

Early in life, Mr. Fixx was an accomplished brainiac. He started running, quit smoking, and became very fit when he was 35; wrote the book and became famous when he was 45; and dropped dead at the age of 52 of a massive heart attack.

If it could be anyone's fault or not that they had a heart attack...this couldn't happen to Mr. Fixx, right? But it did. Where diseases are concerned, there is no way to choose all the variables for everyone in a group like the medical profession tries to do. They would never have guessed that this would happen to Fixx.

And they also don't realize that some folks are so strong in constitution they don't even have to think about it happening no matter what their habits are. We all know a smokin', drinkin', bacon eatin' dude in their 80s. We are all so different that when doctors tell you percentage numbers (and likelihoods and the chance that events will occur later in life), it is not much more reliable than your average casino game. You can trust me; I have done all the math.

Sadly, these days the medical industry has taken things even a bit further down the wrong path. They are actually prescribing very dangerous drugs like cumadin and statins to massive amounts of people in their fifties to (maybe) prevent them from having strokes in their seventies and eighties. These drugs are extremely expensive and you can never stop taking them safely once you start because they change the way your body works.

Their usage is relatively new and there is an equal amount of studies saying they do work and they don't work. Sadly, in my research I found many patients who are on these drugs didn't know they couldn't get off them easily, if ever. They weren't told that before the regimen started and worse, they didn't find it out for themselves.

Why do people take such a course of action when it may or may not work and it is so dangerous? Back to fear. You don't want to take hideous dangerous expensive drugs but you don't want to have a stroke either. But how realistic are your fears?

We have 300,000,000 citizens in the US. The Center for Disease Control says 140,000 of us have a stroke each year. This is roughly one half of 1 person in every 100. And it is estimated that 20,000,000 Americans are currently taking statins to avoid these 140,000 strokes. This doesn't sound like a smart plan since these drugs are so costly and harmful. We can assume that many times the people are taking these drugs than is necessary. If we scale that back to prescribing only in solid cases of impending stroke instead of assuming everyone will have a stroke we could save a ton of money and probably several lives, too.

Again, fear is your enemy here and it won't dissipate if we all keep pointing fingers of blame at patients for their diseases and making them feel guilty for making their own individual decisions.

Testing 1, 2, 3, Infinity

For my entire lifetime, until a couple of years ago, all women and teenage girls were told they MUST have a pap smear every year from puberty on to prevent cancer. The new information says we need to have them first by age 21, then over 30 years old every 3 to 5 years, and after age 60 not at all! Where do I go to get my thousands of dollars back for the unnecessary pap smears and expensive office visits that go with them that I've had throughout the last 50 years?

Unnecessary testing has much the same sort of crazy contradictory math you saw when we discussed statins and strokes. The docs are hoping to stop cervical cancer so they go completely nonsensical when they think they can do it with yearly pap smears and waste is born!

This goes on in nearly every single compartment of the medical business.

When you turn age 50, it gets even more outrageous. They thought they could cure colon cancer if everyone got a regular colonoscopy from age 50 on. This procedure also costs a lot of money, has risks, and is often unnecessary.

Case in point, a patient I talked to said she was 70 years old when she was frightened by her doctor into getting her first ever colonoscopy. She wasn't having any problems. She had the procedure and they found two small polyps which they removed. The very next year, they were pressuring her into having another one. If a person's colon shows no problems and two little polyps that took 70 years to grow, why does the doctor want her to have another one the very next year?

http://blogs.wsj.com/health/2011/05/09/significant-number-of-medicare-patients-getting-too-frequent-colonoscopies/

Money is why. It costs an average of 3000 bucks for this test and an average of 1400 plus bucks are out of pocket for the patient. That's more money down the toilet, more risk to the patients' health, and more damage to the healthcare system.

The next big bullshit story: Mammograms. When I was 35 years old, mammograms were new and my gynecologist pushed yearly mammography on me and tried to scare me with the breast cancer threat. I had no symptoms, no history in my family, and very little cash at the time. I looked into what the procedure entailed. Squeezing the breast flat and shooting harmful x-rays through it every year didn't sound like a good idea to me, so I didn't do it.

When I was about 45, I knew a CEO that worked for a company that made a device that would enhance x-ray films of mammography and was proven to outdo digital mammography in diagnosis reliability. Nobody wanted it because it was not reimbursed by insurance. They didn't care that fewer women would have unnecessary treatments. It didn't matter at all to them--if they couldn't get the money. The company went out of business. I felt even more convinced that I should not ever get any mammograms.

When I was around 50, a panel of experts, appointed by the Federal Department of Health and Human Services announced women should have mammograms only after age 50 and only every 2 years. This is because they found no evidence whatsoever that having mammograms before age 50 is saving anyone at all. In fact, they found tons of unnecessary treatments for women in this age group which likely could have been avoided if---the better mammography film reader was available.

http://www.nytimes.com/2009/11/17/health/17cancer.html

I am so glad I have never gotten a mammogram! This means that had I followed the earlier advice to get a mammogram at age 35 and every year after that until now, I would have gotten about 20 doses of dangerous x-rays shot through my breasts and paid a total price of anywhere from 3000 to 6000 bucks out of pocket FOR NOTHING. Except for this: a good chance at a bad reading leading to more unneeded and dangerous tests and medicine.

This amount times all the women who went through this (I'm baby boomer age) is another enormous amount of money wasted. Someone said to me as a defense in the mammogram argument: "The tests are almost always subsidized nowadays, though." That's even worse. No better way to get patients to adhere to unnecessary treatments than to get somebody else to pay for it. But just like most free things…it turns out to be worth the price.

Just in case you don't believe that the medical community is brainwashing and bullying the public and does not really care about your health, consider this: there are thousands of women aged 35-50 who are STILL today getting yearly mammograms, because they are scared of breast cancer-- and the docs are letting them! This is nuts. Not to mention the extra costs and trauma of unnecessary follow-up procedures on the proven often bad readings on mammograms in this age group.

In addition to cancer screenings that are not nearly as helpful or sensible as the medical community suggests, there are new procedures set in place that I guarantee are going to yield the same shift in opinion over time as the examples above, once the physicians and their affiliated associations realize they've over-reacted again. One is the cholesterol/statin debacle we talked about earlier. Another is rampant diabetes treatment.

Diabetes is a serious illness. It occurs about twice as often as cancer and the medical industry thinks they have a solution for this, too. In the 1980s, the diagnosis of pre-diabetes was brought into practice. Today, the diagnosis is indeed rampant. These are the current criteria:

WHO criteria: fasting plasma glucose level from 6.1 mmol/l (110 mg/dL) to 6.9 mmol/L (125 mg/dL).[4][5]

ADA criteria: fasting plasma glucose level from 5.6 mmol/L (100 mg/dL) to 6.9 mmol/L (125 mg/dL).

What pre-diabetes really means is: you are getting older and you motor needs better care, you are sporting a few extra pounds and need to sleep well, eat a better diet and move around a bit more.

Like so many of you, I was diagnosed pre-diabetic at the age of 45. Without giving me any recommendations on diet and exercise or instructions on how to monitor my blood sugar, the first thing my doctor did was try to put me on Byetta (a serious drug that was brand new) to fix it. That's a diabetes drug. I didn't have diabetes. I fired her.

I learned all I could and fixed my sugar level in four months to well below the normal range with a few diet changes and a little regular exercise. It wasn't that hard. Many years later, I still check my sugar at home once a month and it is fine.

I think this same level of improvement can be accomplished without dangerous drugs in a vast number of cases. While I was grateful for the heads-up, I was disgusted by my doctor's actions. I believe that she did this partly because the drugs she suggested would make more money for her and my improving my health would not.

Worse, she assumed that I was one of the people who are scared to death by anything a doctor says and that I wouldn't take any actions to help myself outside of her ideas. In other words, she treats her patients all the same even though we are all so vastly different.

Now, the medical industry is trying to lower the normal range of blood pressure from the forever held standard of 120/80. These nitwits are now advocating that 120/80 is borderline high blood pressure. This will get more people on more drugs that require regular labs and more frequent visits. More money wasted. More damage to US healthcare. The medical industry is the sole source of damage on these issues because they are testing us to death for no good reason. And they have an accomplished partner in this crime: Healthcare insurance.

The Biggest Scam: Insurance

Health insurance is the most damaging feature of the failing American healthcare system—hands down.

All insurance is borderline con. Ask Floridians or people from Louisiana how reliable homeowners' insurance turned out to be in the last decade or so. And ask them how much more it costs than it did before these companies COULDN'T pay their clients in 2005.

As their punishment, they are awarded complete control and reign over the homeowners; demanding they replace roofs when they are a certain age, or remove recreational equipment from property. Plus, they get to charge a lot more money. Nice work (particularly after a massive fail) if you can get it.

How many of you have been the victims of car accidents where the party at fault has no insurance? (Four times in my family alone.) Why do we have to pay car insurance and the other guy doesn't? Isn't someone supposed to be monitoring this somewhere? I can't get my registration without insurance. How are they doing it? Why isn't the ticket for this as expensive as the dang seatbelt ticket at least?

Speaking of at least, I would think that life insurance would have tipped folks off that insurance is a con. What would make anyone think that a guy, who is insuring that they will not ever die, doesn't have the math of the trick worked out ahead of time? (*see Grim Reaper)

But health insurance is probably the worst insurance of them all. They screw you when you are the most vulnerable. They do it when you are sick. It costs the most, pays the least, and is particularly savvy at dodging responsibility to its customers. It's easy. You send in a claim; they say no. It saves them a ton of cash.

The proof, you ask? In 2011, while the rest of us were slipping into the dismal American economy, health insurance companies made record profits. More than 9.3 billion bucks in the first 3 quarters alone; up 41% from the year before! How does the algebra on that work? I'll tell you again: It's a scam.

It's a pretty damaging scam, at that. Health insurance companies are the ones who are controlling everything in healthcare. You saw what happened to the superior mammography enhancing machine…not reimbursable by insurance? Gone!

They say what drugs you can and cannot take. What tests you can and cannot have. What surgery, counseling, treatment, emergency care, and on and on and on. They and they alone get to choose what is covered and what is not covered. Often they even tell you where you can get these treatments and where you can't.

They dictate all the testing that has to go with everything and can label anything they want as experimental, unnecessary or required.

They try to tell us that the outrageous price of everything is the fault of people who are uninsured and can't pay for their own healthcare. This is not the least bit true.

The proof is in the numbers. The number of uninsured Americans remained relatively level from 1987 to 2008 but the healthcare costs and premiums rose and rose. So, unless the same number of uninsured people were getting a lot sicker than they did in past years; the argument that those unable to pay are driving up the cost is fairly flimsy.

(*Chuck A Sickie: Heal America* by Terry Derwitsch)

The latest (and possibly most successful—hence the huge profits) baloney scheme health insurance came up with is the huge deductible scheme. This is supposed to provide you with catastrophic coverage while you pay for nearly everything but that. Then, guess what? Harvard says you are still one serious illness away from bankruptcy even if you have insurance.

http://www.marketplace.org/topics/life/serious-illness-away-bankruptcy.

An estimated 60% of all bankruptcies in America are due to serious medical issues. A whopping 6 out of 10! A staggering 75% of all of those people are people who actually have health insurance at the time of the bankruptcy! Please read that again and as many times as it takes for your jaw to drop.

Correct. Now they sell you a cheaper plan (cheaper; not cheap) that has you paying everything yourself with your giant deductible AND a 3 out of 4 chance of moving in with relatives if you do get that catastrophic illness. And we all just keep sending the premiums in every month. Like stupid robots.

Scariest of all, your government now wants to force you to have this coverage. This is like throwing gas on a fire. How is forcing everyone to pay these villains going to make everything all better? Tell me how. When we know what we know about how these companies are already treating us poorly, are we really all going to line up for it? Insuring everyone will not fix anything; it will make it much, much worse. Fight this with everything you have.

Sueage

Like so many other areas of our economy, medicine has been more than mildly affected by lawsuits. Everyone sues everyone for everything in medicine. A few decades ago we may have been able to control the damage this would cause to healthcare if we had stopped doing it. But now, it may be too late.

The damage is this: The result of decades of across the board lawsuits in medicine, spurred on mostly by unethical lawyers, has changed medicine completely.

Doctors and hospitals are increasingly blatant with their distaste for patients who aren't their puppets. They scare us to death about diseases we don't even have yet. They test us and drug us to pieces and say it's in fear of lawsuits and in favor of good health. They choose treatments across the board to protect themselves instead of treating each patient in the best way for that patient. They say they have to do this, in order to shield themselves. It's all lies.

That might have been true decades ago, when the sueage first began flowing. But now it's just false. There's no incentive for the medical industry to even fight lawsuits anymore. They pay them like all their other bills. It's become completely meaningless. Except for one thing:

The docs make a lot of easy money doing business this way. Scads of them own their own labs and testing facilities. And some even have their own pharmacies to accommodate more profit. They do this despite the fact that they are clearly not the least bit afraid of lawsuits anymore.

They all have insurance partners for that nowadays. Years ago, they all whined about how the premiums for malpractice insurance and the cost of lawsuits would put them all under. That didn't happen at all. Nowadays, they've adjusted their prices and budgets to pay those malpractice premiums with ease. That's where the REAL cost rising occurred in docs prices and insurance premiums—NOT from poor folks! From waste and from suers and the adjustments the medical business has made because of suers.

They have the powerful malpractice insurance people on their side. Plus, lots of states have limited the damage levels awarded to victims to accommodate the malpractice insurance business. So, what's the risk for them, really?

Back when this all started everyone called the lawyers ambulance chasers and it was an insult to lawyers to be labeled as such. Now they openly chase this stuff down in advertisements on television!

I like the ones the most that directly follow a commercial for drugs with 56 seconds of death warnings that tell us if we took a certain medicine we can sue the pants off of the drug companies. Right, we can if we want to make the lawyers richer. It's just another business. But if we don't see that it is just another business it can cost us our health and tons of cash.

Just like heath insurance is an important business in America because it has done the most to help ruin American healthcare. They've made the most out of your misery, don't condone it. Stop sending them money if you possibly can. And again…fight like mad for your Constitutional right to refuse forced health insurance coverage or this will all get a whole lot worse.

Records Breaking

We've been talking about how doctors treat everyone exactly the same and how stupid this is, as we are no two alike. (There will be more on this in the next essay.) Outside of the obvious detriment here's another way this is going to bite you back.

If you have a break in health insurance coverage, the industry can ban you from coverage based on your previous records and that is exactly what they do. Again they minimize pay-outs by banning those people that they already know need the coverage or might soon need it.

Now here's a warning: If you ever have a procedure, to rule out heart disease, you can be denied coverage by health insurance companies, even if there's no diagnosis. Even if the procedure shows no problems, you will be labeled as a heart patient. This happened to several people in my research. After they were laid-off from their jobs, they were labeled as uninsurable because of this unnecessary testing.

Usually, these patients are diagnosed at a routine check-up with arterial afibrillation. They feel fine. They have no symptoms. This "diagnosis" is not an illness. It is a condition shared by at least 2.2 million Americans. The biggest threat, according to docs, is stroke. They continue on, with the danger story, frightening the hell out of these patients.

Only about 14% of the 140,000 strokes we have per year are associated with this condition. That's about 20,000 strokes per 2,200,000 million diagnosed afib or less than 1%. That is not enough to warrant putting all 2,200,000 through cardioversion, other rigorous heart tests and procedures and putting them on dangerous drugs like cumadin and statins. And I am not the only one who thinks so:

"In a move likely to alter treatment standards in hospitals and doctors' offices nationwide, a group of nine medical specialty boards plans to recommend on Wednesday that doctors perform 45 common tests and procedures less often, and to urge patients to question these services if they are offered. Eight other specialty boards are preparing to follow suit with additional lists of procedures their members should perform far less often."

(http://www.nytimes.com/2012/04/04/health/doctor-panels-urge-fewer-routine-tests.html)

This news came about 6 months too late for the patients I interviewed. One had already gone through cardioversion and cardiac catheterization. He was also told he may have severe blockages and/or melanoma in his lymph nodes. The cardioversion did not work and all the other tests he had revealed that there is nothing wrong with him. He paid several thousands of dollars out-of-pocket to get this information. Thankfully, he wisely did not agree to go on statins or cumadin, as was suggested at the time of these tests (he didn't because his cholesterol is and always had been normal).

So, the only thing he got out of the whole charade was stress, at a time when he was already stressing to the max, a huge bill to pay, and now he can't get healthcare coverage.

In my research I found people who were diagnosed with controlled sleep apnea, pre-diabetes, pre-high blood pressure, and many other non-illnesses that were also denied coverage. What can you do?

It all goes back to choices. Know that whatever goes into your records is going to be contorted by the insurance industry to avoid paying you. Know that doctors are treating and testing us all as one patient without reasonable math to back them up--even according to their own industry leaders and specialist panels. Make your choices accordingly to protect yourselves—and you will probably end up saving a ton of cash in this process.

Hereditary Risks

It's actually funny to list all of the things that are labeled as hereditary. After a few minutes or so, if you're really smart you get the joke. Everything is hereditary. Duh! Wouldn't it have to be since all people share more than 99% of the same DNA? This is somehow a newsflash in the medical community.

They use it as part of a ridiculous premise that you need them more because you have illnesses in your family. Or perhaps you need treatment before anything ever even happens.

So, in a family with four offspring, where two are likely to get ill (however accurate that really is), four are often treated. That's twice as much treatment as necessary. It also costs twice the money. More waste-- even if it really was effective.

It's true that we share mostly the same DNA. This is how the industries convince us they can treat us all the same and it will work. But here's what they don't think of:

We can all learn how huge our DNA bank is.

We can realize that the less than one percent difference among each of us is enormous relative to our makeup. Put simply, the more variables there are, the more differences there will be. Think of a deck of cards.

If each person is a deck of cards and in each deck you change only one card; they are not even close to the same as the other decks anymore. One person can make 5 of a kind aces and another person can't from changing that one card. Some get a better chance of royal flush…others have no chance at all. When you understand this, remember there are only 52 cards in a standard deck. You can probably track those differences well enough to solve some of their problems.

Conversely, the human body contains more than a billion miles of DNA. A billion miles, I say! Do you really think the docs have it all worked out? They cannot possible have done this. The very best computers and thousands of people couldn't do anything but mostly map it in more than 20 years.

Hereditary statistics are part of the harmful practice of listing risk factors. You feel good but you're too fat, you're lazy, you smoke, your family has a history, and/or you don't have the right numbers on your test results.

Risk factors lead to more unnecessary fear and more unnecessary medical treatment. (Again, recall the statins/stroke story.) All the factors they use to monitor everyone in exactly the same way cannot ever work well. We are all too different. It wastes a lot of money that could be better spent on real time, actual healthcare that people need to stay alive today!

For another risk factor malarkey sample, consider the BMI (body mass index) that physicians use to see if you are "at risk" for heart attacks, or diabetes because you're just too fat. Tim Tebow, college football superstar and NFL quarterback, has a BMI of 29.4—borderline obese. Do you really even need any more examples? Have you seen Tim Tebow? He doesn't look the least bit unhealthy to me.

(http://personalliberty.com/2012/05/10/if-you-cant-stand-the-fat-get-out-of-my-kitchen/)

You can see that through manipulation of information and misrepresentation of what's really important to our health, the costs are staggering. You already know insurance is raking it in at a ridiculous rate.

Now, look around your physician's parking lot—what do you see? Beamers and Audis and Jags, I bet. Looks like the patients are the only ones not benefitting greatly from the foolish risk factor driven practices. Do yourselves a big favor. Don't put too much stock in them.

Newfangled Diseases

I pray I'm not the only patient who feels like the medical community is inventing new diseases in a much more lenient way than they use to; so they can make more money. I wish I wasn't nearly all alone in thinking that some of these made-up illnesses are nothing short of evil.

We've already talked about the cholesterol, stroke, diabetes, high blood pressure, money making machine. It's fine if you have done your research and you decide you want to risk using pretreatments as part of your healthcare. I wouldn't try to stop you. You are a grown-up with free will.

I'm not a doctor, as I said before. I'm a patient. I only want you to think for yourself. Oh, and I want you to pay for it yourself, too. Because I know that the total cost of it is truly destroying people. Also, I want you to keep in mind, that kids can't make their own choices.

Having said that, there's one newfangled diagnosis I cannot sit still for any longer (pun intended).

I'm talking about ADD and ADHD. These diagnoses are so common in the U.S.--I know I don't have to spell them out. There isn't anyone left who hasn't heard of these "diseases". There are about 8 million kids taking Ritalin (or worse) for this today. It's even in the spell check!

Here's the history:

When I was a kid there was no such illness. Where did it come from? Turns out it came from school.

In the 50s, 60s, and 70s, America had the best schools and the best students in the world. States and local school boards were running the show. Parents and teachers were as one and commanded respect from kids. We had the highest graduation rate in the world and the highest scoring students in the world. That's why everyone in the world wanted to come here to go to school.

In the 80s, these rankings begin to slip down and continue down for decades until in 2008, the high school graduation rate in most cities is only 50%. I know it's hard to believe, but it's true. Where do we rank in scoring in

2008? 24th. (I'm sure that here in 2012 these figures are probably worse.) The government, with their war on drugs farce, had the people believing kids were too lazy to finish school because they were on drugs. They were drugged all right. They were drugged by their schools.

Here's how I make the jump (only a small hop really) to they got ADD/ADHD from school:

Correlating exactly with the figures above, on a timeline:

50s, 60s, 70s—No such illness, US #1 in Education, No Federal Department of Education, No students medicated.

80s--Federal Government forms the Department of Education and gets involved with schools, offering monetary assistance (via your tax dollars) to schools who comply with certain guidelines, etc. (like the FCATs of today, for example). Why this was done when we already had the best schools in the world is beyond my comprehension. You'll have to get those answers from somewhere else. But my opinion is that it was a sell-out.

The schools needed money and school leaders sold the teachers out, and worse—the students, too. Thinking they could get some easy federal money (your money) school leaders threw the babes out with the bathwater. This is why I am not a teacher. They don't teach kids anymore; they subdue them and I won't have any part of it.

It was in the 80s that massive amounts of kids were being diagnosed with ADD/ADHD. Nearly 500,000 were prescribed powerful drugs to assimilate them to the classroom over the next dozen years or so. This was an extraordinary error. I think we let them get away with it because we trusted them.

We came from a public school system that could be (mostly) trusted. But what we didn't realize was we came from a public school system without the feds in it to mess it up like this. The idiot people that were in charge should have tried to change the classroom—not the children. We have more and more citizens opting to home school and this is one reason why. But I digress…More about this disaster we call American Education in my next book. Let's get back to healthcare waste and danger.

In 1991, the figures make a science-fictiony ten-fold jump to 5 million kids being medicated. When the figures finally come out a few years later, researchers want to know if an event of some sort was tied to this enormous growth of "disease". They found one that would have made George Orwell say I told you so.

In 1991, The Federal Government offered a monetary compensation per student being medicated for ADD/ADHD. They paid the schools for every kid that was medicated. They paid them. The amount of kids on medication rose from half a million to 5 million in a couple of years. And then it grew to about the 7 million we had in 2008 (likely closer to 8 or 9 million now).

After this happens, the graduation rates plummet all over the U.S. to a dismal graduation average of roughly 50% and 23 other countries fly past us in achievement. It's because the kids are drugged to pass an unreasonable social standard and cannot think critically on these powerful drugs.

My heart breaks when a parent says to me, "But, he does so good on the drugs." I know this looks true; but it is similar to saying your electric razor works best when it is not plugged in.

Please, I am begging you. I know that there probably are a handful of kids that can't be helped any other way. But I will never accept that there are 7 to 9 million kids who need to be drugged to get through school and neither should you. If this is the case the SCHOOL needs a cure, not the kids. Help me help them. Assess your other options and get involved in saving your child's very life.

Save them from this massive black hole of being raised on drugs and unable to think on their feet and then dumped out into a world they haven't got the skills to handle. That is; at least not without a lot more drugs.

http://www.youtube.com/watch?v=1KWSIl08XQQ&feature=player_embedded

The Cancer Circus

We have a large number of cancer victims in America. We think this number is bigger because they cram it down our throats all day and night—but still—it's a problem. The treatment is a bigger problem. The tortuous treatment of these folks keeps me awake at night; way worse than any scary freak-show clown ever could.

In my research, I have come across several terrifying facts about cancer treatment. It's really awful. Stories like…"they changed the tubes and I asked why and was told the regular tubes would melt from the medicine they were gonna put in her that day".

Don't get me wrong. Lots of folks are willing to do the treatments and think it has saved them. That is their choice and I say, "Good for them!"

This is just my opinion; I'll remind you of my patient status. I don't believe in what the medical and drug industries are currently offering patients for cancer treatment and apparently, neither do some of these leaders in the business themselves.

One patient I talked to, in 2010, suddenly had her treatments cut in half. This happened as she could still see her tumor. She still needed the treatment they strongly suggested. They just couldn't give it to her because of a shortage. That's right, a shortage.

This particular so-called important life-saving treatment forced on cancer patients as their so-called best shot has platinum in it. And it is just cutoff like the snap of a finger when the price of platinum goes up too high for the pharmaceuticals to turn a profit on the stuff. So, they don't make it. Too bad, go home. Some balls they got. They turned a sizable profit that year on their other poisons and did this anyway. Do you really think it's because they want to cure you? I don't think so. What they want is cash flow.

I've learned volumes and volumes of information about cancer. I'm thinking there are plenty of people who believe they had cancer when in fact it wasn't cancer. Or it may have been the kind of cancer that was not likely to grow fast enough to kill you before your heart or mind gave out anyway.

And I learned that remission counts as a number in the cure ticker, too—even when these folks die next year or a couple years later. In my eyes, they are lying about how many people they can cure.

There are also people (the vast majority) whose cancer could not be helped by anything in reality and they never seem to know it until they have had agonizing and life threatening treatments under their belts. These treatments drain their life savings. (That is if those kingpins decide to make the stuff for them; just saying.)

Incredibly, nowadays they have treatments for cancer that can't save the victims. Yet, they want these victims' relatives to get regular treatments so they can start not curing them sooner? I need clarification on this, I think. Outside of being risky and a bad use of your limited time on this earth and your resources, this is a very wasteful and unnecessary treatment cost overall in the current climate. We have to stop treating people before they become ill if we can't handle (financially) treating the people who are already sick. How can this not be clear? Is it fear, again?

Yes, it is fear and worse. It's money. There's a man named Rick Simpson from Canada who you must investigate. In this particular case, there is safety in numbers, indeed.

This man claimed to have a cure for cancer and many other diseases and tried to get the medical community in Canada to listen to his research. The ultimate response was his arrest and the confiscation of everything that he owned. Keep in mind that Mr. Simpson never made a cent on his wonder drug. He gave it away and taught people how to make it themselves. If you care about cancer at all, I urge you to watch his video, post haste: it's called "Run from the Cure".

http://www.youtube.com/watch?v=0psJhQHk_Gl

Note to Readers: When I was a kid, my parents had a medicine cabinet drug…a tube of goo they called black salve. I could swear this probably had cannabinoids in it, when I look back on how well it worked. Plus, it disappeared from drug stores when I was still young and I always wondered why. I'd love to hear from anyone who knows and I bet Rick would, too.

When he was released from jail, Mr. Simpson fled to Europe and spent the rest of his life savings receiving threats and living in fear for about the last 5 years. He was cleared by the Canadian Government in early 2012. He is reluctant to return home, he's all but penniless now and fears for his very life, every day. What could prompt such a serious reaction from his government and the world?

The answer is hemp. Rick makes oil out of hemp that is a super salve. You can fix things and cure things left and right with this stuff, including cancer. There are dozens of examples of success in his video and Rick is currently visiting numerous radio shows in the states to promote his new book about what he has gone through trying to help us get this medicine and gaining support. These radio shows take in scores of callers who are making the stuff themselves and it is working for them.

So, I'm gonna bet that this is about the point where you say hemp oil can't be a cure for cancer, because if it was, there would be no cancer. And I don't blame you a single bit that you have been brainwashed to think this way. I thought that way myself when I was younger. A lot of rich and powerful people count on that. But you gotta ask yourself first—is there a reason they wouldn't cure cancer with hemp if they could?

And the sad answer is: why yes, yes there is. It is money. No one can patent the hemp oil because it's a plant. They won't make any money on it. So, they won't produce it no matter what it cures. Certainly would not make the money they make on medicines with platinum in them, would they? Do you remember the non-reimbursable mammography enhancer machine's fate? And look--no lawyers anywhere on this scene.

Why do you think lawyers are suing Skechers saying they don't really build leg muscles and turning the other cheek when a citizen has a cancer cure for us that the government won't let him promote? They won't make any money!

Lawyers make millions suing orange juice companies for the "natural" claims on the labels and stupid stuff like that. And they totally ignore the places where we really need them because they won't make any money on those cases. Like I say, it's just another business. Attorney concern for people is a fairytale.

If I am wrong, where are they? Where are the lawyers who will fight to keep our government from illegally searching us at the airport? Have you seen a single one even talk about that being wrong? No, you have not because they're too busy looking for cash cases. Always remember that. If it wasn't true, they could have helped Rick Simpson get us this medicine, too. But they are not helping him.

Nevertheless, there is very good news as I write this on Rick Simpson's story. Tommy Chong, sadly, has been diagnosed with prostate cancer and is openly treating it with hemp oil. I know that plenty of stupid people will say it isn't real when he cures it. He'll be lucky if they don't put him in jail again! Tommy's just a pothead, right?

The great news is they don't realize how many fans he has and that these fans will all rally around him and the cure like bees to a hive! We can work to make them give us this medicine and if they don't we can make it ourselves.

Rick Simpson deserves the Nobel Prize in medicine when the truth comes out. I pray I am alive to see this happen. He gave his very life to get us this oil. Who does that anymore? I urge you to follow him and watch and read everything you can of his and when you become convinced…help him out at this website:

http://phoenixtears.ca/

A portion of the proceeds from the sales of this book will be sent to Rick Simpson, c/o Phoenix Tears, 404 West Easter Avenue, Littleton, CO 80120.

Fake Pretty

I'm sad I even have to write this so I'll keep it short.

A fantastic feat of medicine's progress was accomplished in the form of plastic surgery. People who had horrific accidents and terribly deforming burns could have reconstructive surgery completely renewing their lives. Wonderful!

Now we use a lot of medical resources and take a lot of risks so people who simply don't like their noses or chests or lips or age or whatever can change them. Guess what was accomplished by this brilliant move? The insurance companies win again! And the patients continue to lose.

I met a patient who was denied surgery (that was medically necessary) as cosmetic and it nearly cost her life before they straightened it out and allowed her to have the operation.

Here's a suggestion: How about we teach inner beauty? Why not teach things like love, commitment, discipline, patriotism, pride, self-accomplishment and compassion for others?

This should make us feel pretty; not how big or small our noses or any of our other parts are. Save the surgery for life-threatening situations. Don't treat it like a trip to the candy store.

The Real Drug Problem

This is the most important information in this book. It is the most real. It won't be easy to read. Please take it seriously.

I'll start by saying, I'm absolutely amazed that cannabis, marijuana, pot, herbage; whatever the proper term for the weed is today; is still not allowed in America. It's such a non-threat and in fact most of us know its real value. We are supposed to be civilized and live in such a way that the majority of us want to live. Why aren't we doing that?

I feel certain there is enough support to outlaw these arcane, money-making, and money-wasting laws.

That's about 1% percent of the real drug problem in America today. Pot is still a crime. It's stupid. It's long past time to decriminalize it. It's long past time to embrace it as a super drug. Yawn, get real.

What's the other 99% of the real drug problem?

Opiates.

Not heroin, not opium, not codeine, not even morphine. Those are the natural opiates.

I'm talking about the man-made kind. I'm talking about the methadone, the percocet; their partners: fentanyl, xanax, and oxycodone; and the grand poobah of all painkillers: oxycontin.

Synthetic opiates will kill more people today than all the other drugs combined—legal and illegal—on any list, many times over.

In fact, these drugs will kill more people under 30 years old today than everything else combined (including traffic accidents; which up until now were the main cause of death for this age group for the preceding 60 years at least).

Look at all the studies you can find. The numbers will blow your mind. These numbers don't even contain all the kids whose families didn't share their drug history with anyone out of shame and embarrassment. It doesn't include the number of deaths by other opiates stemming from Oxycontin usage either.

They are huge numbers and it's being swept under the rug by everyone. Here's why: Money, again and the worst case ever of blaming the patient. Here's what happened:

The "experts" all agree that methadone, which was first invented to try to avoid opiate addiction via morphine in a surgical setting, is even more addictive than morphine.

http://www.drugfreeworld.org/drugfacts/painkillers/international-statistics.html

But they keep giving it to you anyway. I mean—they keep selling it to you, anyway, as the cure for heroin addiction and raking in the bucks. In plain English (and in reality), you have a better chance of kicking opiate addiction on morphine than you do on methadone. The government has these poor trapped folks lining up for methadone only to remain opiate addicts every minute of the rest of their lives. And they know they are doing it. They don't care.

Worse, they make money doing it and they are blaming the patients for failures. The addicts' families have been assured that this is a good treatment for their loved ones and if it fails--it's all on the user. The real truth is they set them up to fail by addicting them to something even worse. They cut out the black market middle man and have them addicted to their drug instead. They never get anyone off drugs. They are just transferring where the cash for drugs goes--into their own pockets. How is that a choice for the addicted? Do you get it now?

Now they have an even bigger cash crop for the newly synthetic opiate-addicted. They can get away with it, too, because oxycontin is a death trap drug like no other drug before it ever was. People are literally dropping like flies from this drug. Most of the users who haven't died yet, wish they were dead.

You can get severely addicted to this drug in one usage at a much higher percentage than heroin or crack or any other drug. Once you are addicted, the effects of the drug do not allow you to sensibly choose how much you can take and stay alive. You know this is true with heroin; that's why people mix it with other stuff and die. This artificial stuff is a hundred times worse in this capacity.

This is because after regular use begins, Oxycontin could be destroying your liver and killing you; but its design could make you feel like you ain't even high yet while this is happening.

You start snorting it instead (if you didn't start that way) and even shooting it up. The more you take, the worse it gets. And it is crazy expensive. (Some kids turn to heroin in the process only because it's cheaper on the street.) So, you have to take more to feel high and you have to feel high or feel extremely sick and it finally kills you.

Also, if you stop for some time and go back to using it, your brain has kept the same information as when you were using before, leading lots of people to overdose the first time back. So every relapse has a greater chance of resulting in your death. The drug maker knows this. They still don't care. They blame the patient if they get addicted. They blame the patient when they can't get well.

But oh joy---these legal pushers have a cure for this, too.

It's called suboxone. It's the methadone plan with a different name and a more dangerous drug. It was invented specifically to help overcome synthetic opiate addiction. It got passed through the so-called legal safety channels, (even though it was found to be just as addictive as oxycontin and pretty damaging to some of your organs), on the promise that it would be used sparingly, carefully, and in short courses. It isn't. And it's just as crazy expensive as the Oxycontin is on the street.

There are patients who have been hooked on suboxone for many years. They get it right from their doctors, usually with xanax chasers. Most of them don't even know that suboxone is extremely dangerous for them and that they are only supposed to take it for a week. They usually find out in detox; if they are lucky. If not, their loved ones find out in the morgue. And nobody cares. It's their choice, right? They're just junkies, right? They didn't fight hard enough, right?

Wrong. Wrong. Wrong. They are your sons and daughters, your moms and uncles and spouses and lifelong friends. Think and be honest…I know you know someone. These are all people under 30—I ask you: Who will work and support the country in the next few decades if scores of them are dead? This scenario is a number so big you can't imagine it. Two whole generations, are struggling in a death maze created by pharmaceuticals. A maze where there is no way out, and nobody is even talking about helping them; everyone is blaming them.

How was this maze formed? Oxycontin was invented to be a better pain reliever than any other drug invented before it and it certainly is that. And the two newest generations of young adults know it. But they think it's a safe drug because stupid doctors give it out so readily and because of the way they were raised.

All their lives, we've been pumping medicine into people 30 years old and younger. They trust man-made medicine even worse than the millions of older folks who are on statins and don't know that they can never get off of them safely and that they probably won't make a difference for most of them anyway. This has done severe damage to healthcare and caused this enormous drug epidemic that the government (and indeed, the world) is now ignoring.

It's especially ridiculous that this is happening when a few cases of SARS virus makes the whole world go insane with worry. We sell out of face masks and gas masks in a couple of days. Oh, and did we need those? No, we did not. Some of the vaccines for that turned out to be bad or counterfeit even--from the haste makes waste principle--making a bad triage choice even worse. How much cash down the toilet on all of that?

At the same time, we have thousands and thousands of people under 30 that are dying at the hands of pharmaceuticals' fake heroin and nothing. No attention from anywhere official. It's disgusting; if not criminal. In a country that claims to care so much for young folks, it's blatantly shameful.

Other ill effects of the drug age are surfacing as we speak. We have managed to completely destroy antibiotics by pumping everyone full of it as children for simple things like ear infections. When I was a kid, short of broken bones, my family doctor instructed us to relieve the pain, stay in bed and ride it out. We all got a couple of ear infections each (maybe) throughout our childhoods. We built our defense against ear infections and everything else from the inside out, making our bodies stronger, like you are supposed to do.

Kids now are inoculated every five minutes or so. They have taken tons of antibiotics, even for a cold because they can't miss school to rest and fight it off, they way they should. So they won't get another cold.

They are all doused in anti-bacterials, killing even all the good bacteria. (Yes, there is really good bacteria.) Soap was good enough for our immunities for over a hundred years. Now they want them really clean, all the time and it's making them weak. Google MRSA for the worst thing we've made happen from abusing usage of antibiotics and our need to be ultra-germ free.

If we claim we don't even want a germ on our kids in this country, how in hell are we ignoring the biggest health problem they have ever faced? One that was mostly created by the drug companies and is being ignored and perhaps even perpetuated by these very same companies?

How do we dare to call these victims useless junkies and turn our backs on them when we don't even know what they are really up against? I demand to know how. I also demand that it changes right now.

Afterward

If you didn't notice the many times I told you in this book that I am a patient…I am not a doctor, please notice it now. This book is not a medical book. It is a book designed to train you to think critically about medical issues like I want you to do in every facet of your life.

It should be obvious to you now that the REAL problems in healthcare are not going to be solved by anyone but the people. Let's get solid on this. Let's be realistic. You see the way the industries currently running healthcare actually care about your health. In so far as it will fill their accounts with real cash—nothing more. YOU and I have to care. YOU and I have to fix it.

It's time for the people to shape healthcare. Not the docs, not the kingpins, not the scam-artist insurance industry and most certainly not the already oversized, over-bearing, Federal Government with the poor performance record. I'm talking about: YOU and ME. We are the ones who pay for all of this mess. Let's clean it up.

www.ingramcontent.com/pod-product-compliance
Lightning Source LLC
Chambersburg PA
CBHW021904170526
45157CB00005B/1965